MARVELS
OF THE
MOLECULE

MARVELS OF THE MOLECULE

Lionel Salem
Université de Paris-Sud, Orsay, France

Illustrated by **Colin Rattray**

Foreword by **Roald Hoffmann**
Cornell University

Translated and Edited by
James D. Wuest
Université de Montréal

with contributions by
J. J. Lagowski
University of Texas, Austin

VCH

Lionel Salem
Laboratoire de Chimie Théoretique, associé au CNRS
Centre d'Orsay
91405 Orsay Cedex
France

Library of Congress Cataloging-in-Publication Data

Salem, Lionel.
  Marvels of the molecule.

  Translation of: Molécule, la merveilleuse.
  1. Molecules. 2. Chemical reactions. I. Lagowski,
J. J. II. Title.
QD461.S2813 1987   541.2'2   86-32596
ISBN 0-89573-345-5

© 1987 VCH Publishers, Inc.

Printed in the United States of America.

ISBN 0-89573-345-5 VCH Publishers
ISBN 3-527-26530-9 VCH Verlagsgesellschaft

Distributed in North America by:
VCH Publishers, Inc.
220 East 23rd Street, Suite 909
New York, New York 10010

Distributed Worldwide by:
VCH Verlagsgesellschaft mbH
P.O. Box 1260/1280
D-6940 Weinheim
Federal Republic of Germany

To chemists and non-chemists alike

Foreword

If there were no molecules, there would be no toucan feathers, no perfume of lily of the valley, no Tricolor, no paper nor ink for the score of a Tschaikovsky opera, no life at all. A silly conceit that, of chemists, men and women whose life is devoted to the study of molecules. Yet there is truth to it. All that makes life beautiful, all that makes life go on when it is not beautiful, has a molecular basis. Biological molecules and unnatural creations of the flask alike share a three-dimensional architectonic complexity. The flexible mortar that holds these assemblages of atoms together and controls the ordered ways in which molecules transform is the chemical bond.

As one climbs a hierarchy of complexity from nuclei and electrons to atoms and molecules, one needs the language of quantum mechanics, of waves. That mathematically concise language provides the rules that govern the ways in which electrons bind atoms to each other. Lionel Salem, in this marvelous little book, has provided us with a modern Baedeker, a guide to the seemingly strange country of the chemists. First he surveys the architecture of molecules, sitting still and in ceaseless motion. Then he analyzes the bond between atoms. Translating jargon of the trade into plain English, he shows us that simple concepts govern the structure, motions, and transformations of molecules.

This book is not a paean to technological progress, nor a dirge on its impact on the environment. Instead it provides a harmonious guide with the help of which lay readers may enter the world of a modern chemist. In it they will find wonder and exultation, as William Blake describes it:

> *To see a world in a grain of sand,*
> *And a heaven in a wild flower;*
> *Hold infinity in the palm of your hand,*
> *And eternity in an hour.*

Roald Hoffmann

Preface

Ah! la science ne va pas assez vite pour nous!
A. Rimbaud, *Une Saison en Enfer*

If you are interested in the world around us, you probably want to
know how matter is put together. To understand, however, you
must face the discouragingly difficult and highly specialized language of chemistry. This language is sometimes the refuge of scholars who are not really interested in communicating what they have
learned. Fortunately, though, the logic of these scholars is as simple
as the logic of children.

This book is an attempt to demystify modern chemistry. I want to
show that it can be described in everyday language. Except for the
names of elementary particles, I have intentionally refrained from
using any purely scientific terms. Instead, I have chosen to use
equivalent expressions from ordinary language.

The theme of this essay is the molecule. Composed of atoms just as
words are composed of letters of the alphabet, molecules have
extraordinary variety. There are at least as many different molecules as there are different words in the dictionary. Almost every-
thing around us, including living things, houses, furniture, food,
and clothing, is composed of molecules.

Like objects of everyday life, molecules have well-defined shapes.
It is customary to describe their shapes by comparing them with
common objects: one molecule looks like a chair, another like a cube,
and another like a ring. You should try to think of molecules as tiny
objects that can be touched, probed, or set on a table so that their
geometry can be admired from different directions. Chemists themselves build plastic or metal models of molecules to help them think.

Moreover, molecules act like *animate* objects. They are not alive in
the normal sense, but thermal energy lets them fidget and move
around, and even gives them the astonishing ability to change into
something else. This is a *reaction*: molecules meet and become different molecules. You will be able to marvel at this process without

having to believe in magic, once you have seen how the encounter of molecules simply lets certain atoms move.

Of course, real molecules are so small that billions and billions are needed to make a speck visible to the naked eye. We won't let that stop us! Let's enter this world of the infinitesimal.

Lionel Salem

Contents

1
Form and Fancy

THE GEOMETRY OF MOLECULES

1

THE WATER MOLECULE

Take a drop of water. Measure its diameter. Divide this distance by a hundred, then by a thousand, then again by a thousand. The result is approximately the length of a molecule of water. Roughly a million million million million of these water molecules are needed to form a drop, much like many bricks are needed to build a house.

The water molecule looks like a peach to which two apricots have been attached. The peach in the middle is an atom of oxygen. Each small apricot next to it is an atom of hydrogen. The oxygen atom is joined to each hydrogen atom by a refined connection called a bond. A bond is indicated by a line joining the centers of the circles that represent the atoms. Sometimes molecules are drawn without using these circular contours, and the lines are simply drawn between the centers of the atoms. Two or three parallel lines between the centers indicate that the atoms are connected by two or three bonds. These are the ways that chemists draw molecules.

The nature of bonds has fascinated scientists for centuries. We will study the details later. The bonds in the water molecule are so strong that a drop would have to be heated to over 2000°C in a special furnace in order to break the molecules into their component atoms.

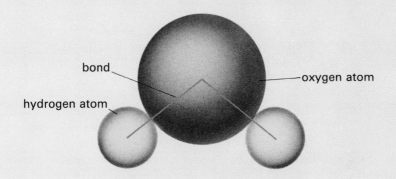

bond

hydrogen atom

oxygen atom

The water molecule resembles a peach (oxygen atom) to which two apricots have been attached (hydrogen atoms).

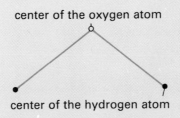

center of the oxygen atom

center of the hydrogen atom

The water molecule can be represented more simply by drawing lines between the centers of the atoms.

2

MOLECULAR BALLET

Separate a water molecule from the drop, perhaps by capturing one that evaporates into the air. Like a top, it spins around at a fantastic speed. In a single second, it completes a million billion turns. Unlike a toy top, the molecular top can spin in three different ways, which are illustrated in the drawing.

While the molecule is spinning rapidly, its atoms take part in an endless dance that is a thousand times slower, but still very fast. They move apart a little, then come closer, greet briefly, and separate again. The bonds between the oxygen and hydrogen atoms stretch and relax like springs, and the angle between them opens up and closes endlessly. Thus the molecule undergoes constant distortions, all with a preordained harmony. The three atoms are joined, and their movements are marvelously coordinated.

The atoms can dance three different ballets: in one, the two bonds stretch in unison; in another, one stretches and the other contracts; and in the third, the molecule does the splits.

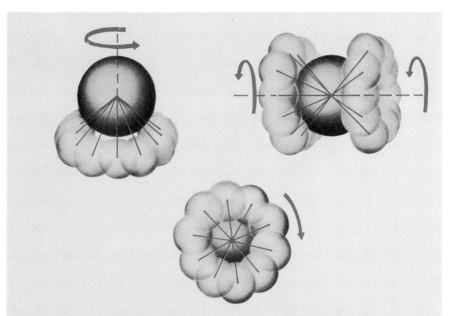

The water molecule is a top that can spin in three different ways.

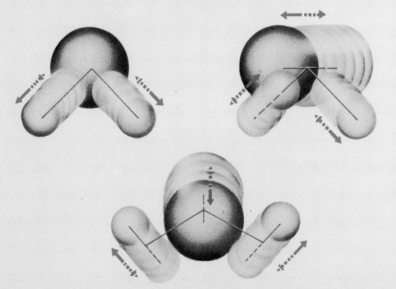

The atoms in the water molecule can dance three different ballets: in one, the two bonds stretch in unison; in another, one stretches and the other contracts; and in the third, the molecule does the splits.

3 MOLECULAR TINKERTOYS

One of the most useful atoms for building molecules is carbon. This is because it can attach itself to four other atoms by forming four separate bonds, all very strong. As a result, nature lets us construct three-dimensional structures by using building blocks of carbon atoms. Molecules of the organic world, like those in our bodies, in green plants, and in crude petroleum, are all built using frameworks containing carbon atoms.

In this family of organic molecules, the simplest member consists of a single atom of carbon linked to four atoms of hydrogen. This molecule is methane, which has the shape of a perfect tetrahedron. Frameworks of carbon atoms can be arranged in other shapes as well, including chains and more elaborate structures like prisms and cubes.

The boron atom can surround itself with more neighbors than carbon can, so it is even more versatile. It forms some marvelously intricate structures. For example, the para-carborane molecule consists of ten boron atoms and two carbon atoms in the shape of an icosahedron, a solid with twelve corners.

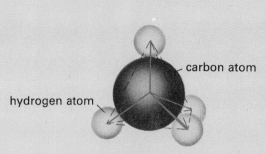

In the methane molecule, four hydrogen atoms define the corners of a tetrahedron surrounding the central carbon atom.

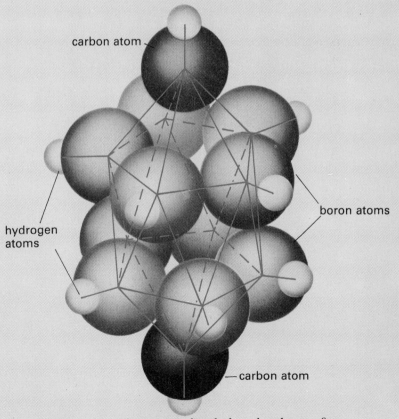

The para-carborane molecule has the shape of an icosahedron.

4

SANDWICH MOLECULES

Molecules of other shapes can be built around metal atoms like chromium, iron, and nickel. These metal atoms tend to surround themselves with a large number of neighboring atoms, sometimes four in a plane, more often six in the shape of an octahedron as in chromium hexacarbonyl. The number of neighbors and their orientation depend on the metal atom in the middle.

Some small molecules, including a few that are unstable by themselves, can latch onto a metal atom and form a stable structure. For example, two molecules of cyclopentadienyl, which consists of a pentagon of carbon atoms, can anchor on opposite sides of an iron atom. This creates ferrocene, a supermolecule that looks like a sandwich. The microscopic world can therefore resemble the world of everyday objects, although it often surpasses our wildest imagination.

Nature even makes club sandwiches, like the tris(cyclopentadienyl)dinickel molecule, in which three pentagons of carbon atoms and two nickel atoms are stacked one on top of another.

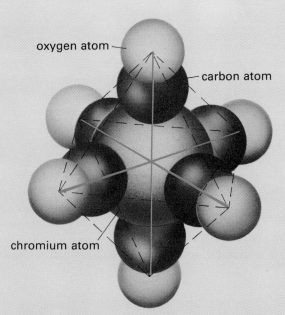

oxygen atom

carbon atom

chromium atom

The chromium hexacarbonyl molecule has the shape of an octahedron.

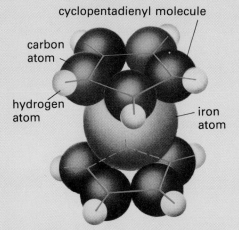

cyclopentadienyl molecule

carbon atom

hydrogen atom

iron atom

The ferrocene supermolecule looks like a sandwich.

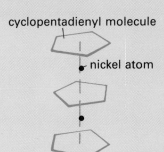

cyclopentadienyl molecule

nickel atom

The tris(cyclopentadienyl)-dinickel molecule is a club sandwich. (This molecule is missing a negatively charged electric particle, so it carries a positive charge).

5

LEFT-HANDED AND RIGHT-HANDED MOLECULES

Here are two living room lamps. They look identical. Both stand on three feet, one black, one light blue, and one dark blue. But try to make them coincide exactly. It's impossible! If you super-impose the two black feet, a light blue one will be standing on a dark blue one, and a dark on a light. Just like our two hands, the lamps are similar but each has its own identity. We can call one our "left-handed" lamp and the other our "right-handed" lamp.

Like these lamps, there are many pairs of left-handed and right-handed molecules that cannot be made to coincide. Amino acid molecules, for example, which are the building blocks of protein molecules, can exist in left-handed and right-handed versions. This is illustrated by the alanine molecule. It is ex-traordinarily interesting that in our bodies and those of other animals only the left-handed forms of these molecules are found. The origin of this preference is still a mystery. Few other natural phenomena have a favored handedness, or sense of di-rection. The earth rotates around its axis in one way and not the other, and the trade winds blow in one direction only, but it is hard to see how these preferences could have influenced the choice of amino acid molecules at the moment when life began.

How would it be possible to separate a mixture of left-handed and right-handed molecules? It is no harder than finding a right-handed glove in a dark drawer that contains many gloves. The best thing to do is reach in and try each glove on your right hand. In the same way, only a molecule with left or right-hand-edness of its own will be able, by chemical reactions, to distin-guish and separate the molecules of a mixture.

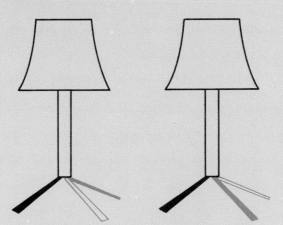

These lamps are similar but different.

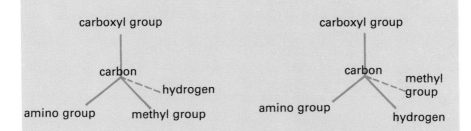

Left-handed alanine molecule.   Right-handed alanine molecule.

In these two molecules, the groups of atoms attached to the central carbon atom have been represented in a simplified way. The methyl group is made up of a carbon atom linked to three hydrogen atoms; the amino group consists of a nitrogen atom joined to two hydrogen atoms; and the carboxyl group is made up of a carbon atom linked to two oxygen atoms, one of which is further connected to a hydrogen atom.

6

MOLECULAR TRAPS

Many new molecules can be built by modifying molecules that already exist. Building these new molecules requires the skills of architects and engineers, because the atoms must be arranged according to a specific plan to produce a molecule with particular properties.

Extraordinary molecules have been created in this way. For example, the cryptate molecules, which are built like barrels with three staves, have been designed to capture atoms and small molecules. The loops of atoms that form each stave contain oxygen atoms, which have a special affinity for positive sodium ions, atoms that have lost an electron*. When one of these sodium ions is locked in the jaws of the cryptate, it can barely escape.

Because the atoms making up the outside surface of cryptates are like those found in membranes, cryptates can pass easily through membranes and enter cells in the human body. Cryptates can therefore be used to carry small atoms or small molecules into cells. Introducing drugs into the blood and tissues in this way may become a powerful tool in medicine.

*An ion is like a normal atom in every way but one: it carries an electric charge, whereas a normal atom does not. Like the terminals of a battery, this charge can be positive or negative. It is positive when the atom has lost an electron, a negatively charged particle; and it is negative when the atom has gained an electron. Opposite charges attract each other, so oxygen atoms, which are rich in negative charge, attract positive sodium ions, which are poor.

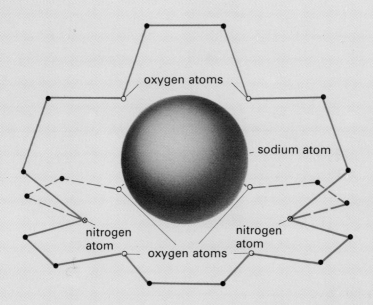

Sodium ions can be captured by cryptate molecules. This drawing shows that the skeleton of the cryptate molecule, which is mainly composed of carbon atoms, places six oxygen atoms close to the captured ion.

7

THE SMELL OF MOLECULES

The benzene molecule is a perfect hexagon of carbon atoms. Each carbon atom is bonded to the two carbon atoms on either side and to one hydrogen atom. These hexagonal units can be used to build a whole family of molecules called *aromatic*. In addition to many interesting properties like great stability, these compounds have characteristically strong smells ranging from the aroma of anise to the stench of mothballs and tar.

Many molecules have a specific odor. Changing a single atom in a molecule may change its smell completely. The hydrogen sulfide molecule, for example, looks almost like a twin of the water molecule, but it has the disgusting stench of rotten eggs. Fortunately, pleasant smells are common. For example, the amyl acetate molecule has the aroma of pears, which can be detected in some brands of nail polish. Why a molecule smells the way it does is a mystery that is still not completely solved, but it depends on how the molecule fits into the cells of our nose.

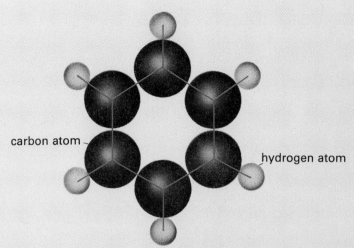

carbon atom

hydrogen atom

The benzene molecule.

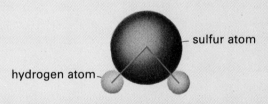

— sulfur atom

hydrogen atom

The hydrogen sulfide molecule (stench of rotten eggs).

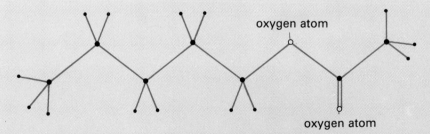

oxygen atom

oxygen atom

Skeleton of the amyl acetate molecule (aroma of pears).
The backbone consists mainly of carbon atoms linked to
hydrogen atoms.

8

BAD MOLECULES AND GOOD MOLECULES

Some molecules, even the smallest ones, are deadly. They can enter our body, and our vital organs can misidentify them. For example, if we inhale carbon monoxide molecules, our lungs believe that we are breathing oxygen molecules. This is possible because both molecules have the same size and shape. But the carbon monoxide molecules will kill in seconds by sticking permanently to molecules in the blood that are normally used to carry vital molecules of oxygen.

Other molecules, like those of plastics and adhesives, have improved our lives. Beneficial molecules have allowed medicine to take great leaps forward. Who has never taken aspirin? Who has never been anesthetized by molecules like ether or novocain? Ether molecules penetrate temporarily into the membranes of nerve cells and prevent them from working. When we breathe these molecules, we lose consciousness. When molecules of novocain are injected, they interfere with the movement of ions and keep nerve cells from carrying the electrical signals that make up nervous impulses. Since pain can no longer reach the brain, these and many other molecules soothe and relieve.

Carbon monoxide molecule (one oxygen atom bonded to one carbon atom)

Oxygen molecule (two oxygen atoms bonded to each other)

A bad molecule and a good molecule: the carbon monoxide molecule is a deadly poison because it sticks to molecules in the blood that are normally used to carry vital molecules of oxygen.

Sketch of the aspirin molecule.

Sketch of the novocain molecule.

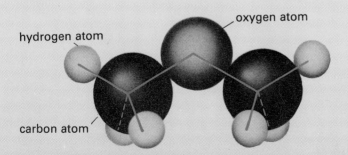

An ether molecule

2
Molecular Assemblies

THE MATERIALS AROUND US

1

A LARGE AND WELL-PACKED SUITCASE

Touch a copper wire or a lump of sugar. What is the relationship between the wire and individual atoms of copper, or between the lump and individual molecules of sucrose?

Anyone who has traveled knows that a well-packed suitcase holds more than a messy one. Whether the contents are baseballs, round like atoms of copper, or footballs, oval like the flattish molecules of sucrose, a suitcase that is carefully packed carries more. Even nature is familiar with this rule, and carefully packs copper atom next to copper atom, millions and millions and millions of them. This magnificent suitcase full of copper atoms appears to us as the piece of copper wire. The wire is therefore an aggregate of countless atoms. In the same way, the lump of sugar is an aggregate of countless identical molecules of sucrose.

It is the huge number of atoms or molecules that makes the copper wire or the lump of sugar visible to the naked eye. This is true of all common materials, including wood, paper, fabric, and metal. Whenever you scrape a tiny speck of sugar from a lump, remember that you are removing a fantastic number of molecules.

2

DIAMOND AND GLASS

Admire the diamond on a woman's finger. Where does the extraordinary hardness of this stone come from? It results from the packing of its atoms, which are joined tightly together by a large number of bonds. Carbon atoms are packed so that each is surrounded by four neighbors, producing interconnected tetrahedra of carbon atoms that are geometrically similar to the methane molecule. The resulting solid is the marvelous stone we call diamond. It is endowed with a hardness that stands up to all tests. When you touch a diamond you should be impressed, for you are touching an individual gigantic molecule. If you try to scrape it, you will not remove a single atom.

The properties of solids formed when atoms are packed together depend on the nature of the atoms and the number of bonds they can form. Take the diamond framework, replace each atom of carbon with an atom of silicon, and then surround each silicon atom with four oxygen atoms. Each oxygen atom can be connected to only two silicon atoms, so the resulting structure is much less rigid and compact than the diamond framework. It contains rings of variable size formed from atoms of silicon and oxygen. Even if this suitcase of silicon and oxygen atoms is packed very slowly and carefully, the result is still something much less ordered than diamond. We get quartz! If the packing is rapid and careless, the result is merely glass.

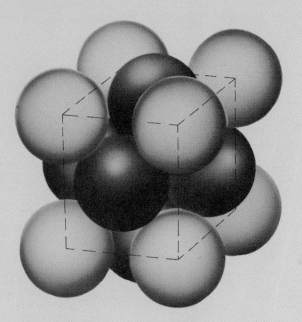

Copper atoms in copper wire are packed in a repeating pattern that consists of eight atoms in the shape of a cube, with six other atoms at the centers of the six faces.

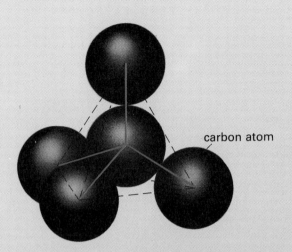

The basic structural unit in diamond consists of a carbon atom joined to four other carbon atoms at the corners of a tetrahedron.

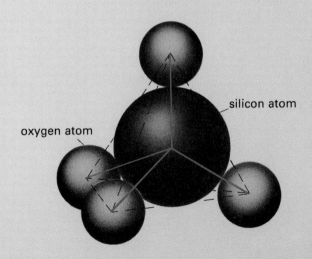

The basic structural unit in glass consists of a silicon atom joined to four oxygen atoms at the corners of a tetrahedron. However, each oxygen atom is connected to only two silicon atoms, one of which is shown here.

3

BRIDGES OVER TROUBLED WATER

It is harder to pack a tidy suitcase with molecules built primarily of oxygen atoms than with those formed from carbon or silicon atoms, since oxygen atoms form only two bonds instead of four. In packing water molecules together, however, nature performs yet another miracle of construction by managing to surround each molecule with four neighbors. This marvel is the result of a special property of hydrogen atoms bonded to oxygen atoms. Such hydrogen atoms can reach out to the oxygen atom of an adjacent molecule to form an additional bond that is weak, but still strong enough to serve as a useful bridge between two oxygen atoms. In this way, each water molecule is linked to four other water molecules.

An ice cube contains countless molecules of water held together by these hydrogen bridges, but the structure is not very stable. The contents of this particular suitcase are fragile. When a little heat is applied, the hydrogen bridges begin to break. Some of the water molecules begin to fidget, then break free. We see the ice melt. A minor event at the molecular level, the rupture of a few weak hydrogen bridges, corresponds to a major event in our lives, the melting of snow and ice as spring arrives.

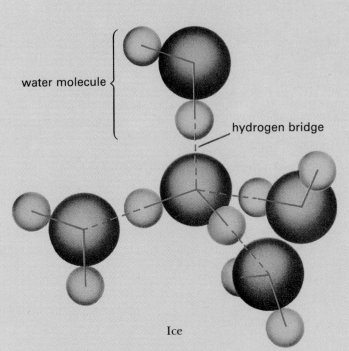

water molecule

hydrogen bridge

Ice

Each water molecule is linked to four neighboring
molecules by hydrogen bridges.

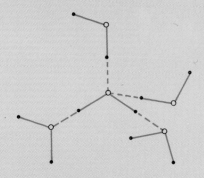

Ice

Simplified drawing of water molecules joined by
hydrogen bridges (dotted lines).

4

THE HUSTLE AND BUSTLE OF LIQUIDS

Unlike water molecules, methanol molecules have only one hydrogen atom directly bonded to an oxygen atom, so they can form hydrogen bridges with only two neighboring molecules. The situation is even worse for ether molecules. In this case, no hydrogen atom is attached to the central oxygen atom, so hydrogen bridging is impossible. When the number of hydrogen bridges is small, the packing of molecules becomes very sloppy. The molecules fidget, move around, and change places restlessly. The result is a liquid. Even at the freezing point of water, methanol and ether remain liquid.

Interactions between ether molecules are very weak. They are merely attractions between regions of one molecule that are rich in electric charge with regions of another molecule that are poor. As a result, molecules of ether can escape from the liquid and evaporate easily. Only when energy is removed by strong cooling are the molecules brought to a standstill. Under these extreme conditions, even methanol and ether can be solidified.

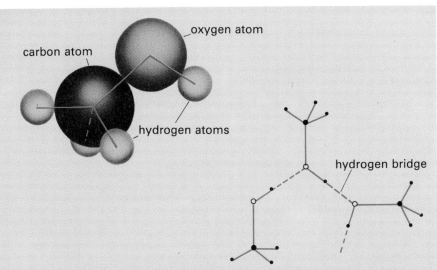

The methanol molecule and its packing in liquid methanol.
A few hydrogen bridges are still possible.

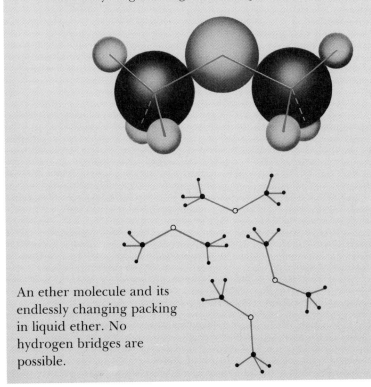

An ether molecule and its
endlessly changing packing
in liquid ether. No
hydrogen bridges are
possible.

5

MOLECULES THAT COUNT AND TELL TIME

Certain fluids are not as perfectly ordered as solids, yet not quite as disordered as the liquids we have just considered. These hybrids are therefore called *liquid crystals.* For example, molecules of N-(p-methoxybenzylidene)-p-butylaniline (MBBA) form a liquid crystal by lining up in parallel like a school of fish.

It is, however, a school of very special fish. When the liquid crystal is placed in a strong electric field, the fish-molecules are set in motion and swirl around in eddies. Under normal conditions, the liquid crystal is as clear as water, but in the state of electric agitation it becomes turbid and diffuses light like a swirl of muddy water or a cloud of dust in a sunbeam.

Molecules of MBBA can therefore be used to make the screens on which pocket calculators display results or digital watches show the time. The screens are filled with a thin film of the liquid crystal. Transparent electrodes, in the shapes of the numbers 0–9, are placed beneath the film. Obeying signals from the calculator or watch, these electrodes subject the liquid crystal to an electric field. Opacity created near the electrode that is switched on makes a number appear on the screen. More recently, light has also been used to create visible displays by causing local turbulence in schools of fish-molecules.

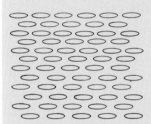

oxygen atom

nitrogen atom

The backbone of the MBBA molecule is primarily made up of carbon atoms surrounded by hydrogen atoms.

MBBA molecules lined up like a school of fish produce a liquid crystal that is transparent to the naked eye.

In a strong electric field, the same molecules swirl around in eddies, and the liquid crystal becomes cloudy.

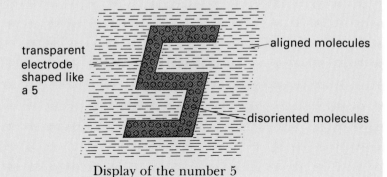

transparent electrode shaped like a 5

aligned molecules

disoriented molecules

Display of the number 5

6

MOLECULES THAT CLEAN

Let's wash our hands with soap and water. How does soap remove dirt so easily? Stearic acid molecules make soap work. Let's look first at a single molecule of this acid. We see a long tail of carbon and hydrogen atoms, and a head containing two oxygen atoms, one of which is further connected to a hydrogen atom. In soaps, the hydrogen atom of the head is replaced by another atom like sodium. If we add a few soap molecules to a large number of water molecules, which is what happens when we stick a bar of soap in water, the soap molecules cluster together to form tiny blobs with the tails in the center and the heads on the outside. The heads point out toward the water molecules, which are relatively disorganized.

When a foreign molecule comes along, perhaps a molecule in a speck of dirt on our hands, it is swallowed up by a blob of soap molecules. This is because the foreign molecule, usually an organic material with a chain of carbon atoms, is attracted by the similar, friendly environment of the carbon tails of the soap molecules. The speck of dirt therefore dissolves inside the tiny blobs of soap molecules.

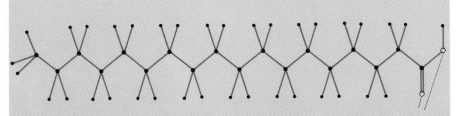

oxygen atoms

The backbone of the stearic acid molecule consists exclusively
of carbon atoms surrounded by hydrogen atoms, and the head
includes two oxygen atoms. In soaps, the hydrogen atom of the
head is replaced by another atom like sodium.

foreign molecules

In this drawing of soap, it is possible to recognize the long
backbones of stearic acid molecules and the heads that point
outward to V-shaped water molecules. The blobs of soap easily
dissolve foreign molecules.

7

RUBBER, NYLON, AND WOOD

Polymer molecules are a lot longer than even molecules of MBBA or stearic acid. In polymer molecules, groups of atoms are linked together in chains so long that they are sometimes visible to the naked eye. The polyisoprene molecule, for example, is made up of thousands of repeating units of a simple group of atoms. Even more impressive is the way these chains of polyisoprene molecules tangle together to form natural rubber.

Polyisoprene molecules love disorder. In rubber, each molecule is curled up like a ball of string. Don't bother to try to straighten things out! If you pull on the molecules, they will grudgingly straighten out; but when you let go, they will tangle up again. This explains the marvelous elastic properties of rubber, which does not break and always snaps back to its original shape.

Chemists have been able to create many polymer molecules that nature had not thought of. These unnatural molecules form many of the materials of everyday life. A familiar example is nylon, which is made up of a tangled mass of nylon molecules.

But let's come back to nature and look through a microscope at a very thin slice of wood. We see cells, and inside them are many different polymer molecules. Of special importance are the molecules of cellulose, which are made up of rings of carbon and oxygen atoms linked by atoms of oxygen into long chains. These are the basic, invisible components of wood from our forests. Our powerful molecular vision lets us trace the beauty of an antique cabinet back to its atomic origin.

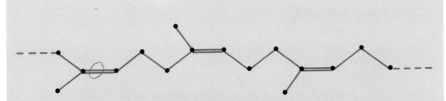

The skeleton of carbon atoms in the polyisoprene molecule has a basic pattern that repeats indefinitely.

A tangled molecule of polyisoprene.

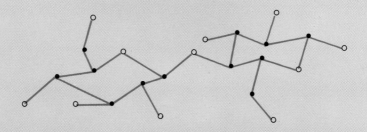

The skeleton of the nylon molecule consists primarily of carbon atoms, with a few atoms of nitrogen and oxygen.

The skeleton of the cellulose molecule is made up of carbon and oxygen atoms.

8

THE AIR WE BREATHE

Interactions between very simple molecules are often very weak. The formation of chemical bonds or hydrogen bridges may be impossible, or the opposite electric charges that attract one molecule to another may be absent. As a result, simple molecules like nitrogen, with two nitrogen atoms triply bonded, or oxygen, with two oxygen atoms doubly bonded, cannot really stick together. At normal temperatures and pressures, these simple molecules remain far apart and exist as gases.

On earth, an abundant mixture of nitrogen molecules (80% of the total) and oxygen molecules (20%) makes up most of the air we breathe. Here and there are rarer components, like atoms of argon. Unlike solids and liquids, gases like air are basically empty space. The distance from one molecule to its nearest neighbor is about fifty times the size of the molecule itself. In addition, gaseous molecules are always zipping around at high speed.

**nitrogen molecule
(two nitrogen atoms)**

**oxygen molecule
(two oxygen atoms)**

argon atom

Molecules in air

Sketch of the air we breathe

3
Stationary Waves

ELECTRONS INSIDE THE ATOM

1

STATIONARY ELECTRON WAVES: WHY ATOMS ACT THE WAY THEY DO

Let's go back to our picture of atoms as fruits and cut one open. The stone in the middle, the nucleus of the atom, is quite small. Although it is only about one hundred thousandth of the size of the whole fruit, it is much heavier than the pulp that surrounds it. However, it is the extraordinary pulp that determines nearly all the properties of the atom.

The pulp is made up of electrons. Nothing quite like an electron exists in the world of everyday objects, since it is two things at once. It acts like a tiny particle that zips throughout the atom, passing through some areas more frequently than through others. But the French physicist de Broglie has shown that it also acts like a wave. This is how we will represent it, although it is a special kind of wave that doesn't really move. To make waves like these, shake a rope that is tied at one end. Normally the waves appear to move up and down the length of the rope, but they can be made to stand in one place if the rope is shaken just right. Like all waves, these stationary waves have crests (dark blue) and troughs (light blue) separated by zones of calm. The vertical distance between crests and troughs is a measure of the strength, or amplitude, of the wave.

The rope waves stretch out in one dimension, along the length of the rope, but waves can also be two-dimensional. Sea waves, for example, ripple over the entire surface of the water, spreading out in both horizontal dimensions. The vertical dimension measures the height of the waves. Other waves, including stationary electron waves, are three-dimensional, and are best represented by the shape of the space that they fill up. We will see that there are many possible shapes. To show crests and troughs in these three-dimensional waves, we would need a fourth dimension. It is easier to use color: dark blue for the crests and light blue for the troughs.

It is important to remember that electron waves stand still, unlike sea waves. They have a definite, fixed, stationary form that is the result of a delicate balance of natural forces inside the atom. These stationary electron waves control the behavior of atoms. Their shape helps determine how one atom can be linked to another, and the pattern of crests and troughs provides instructions for atoms in the process of forming bonds.

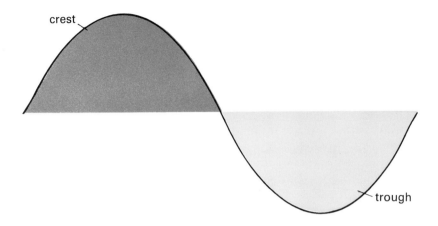

A rope wave stretches out in one dimension, and the vertical distance between crests and troughs measures the amplitude of the wave.

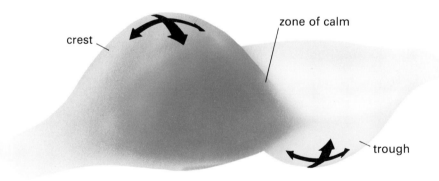

Sea waves spread out in both horizontal dimensions, and the vertical dimension again measures the height of the waves.

2

SPHERICAL WAVES

The simplest stationary electron waves are spherical. At points a given distance from the nucleus in any direction, a spherical wave has the same amplitude. Spherical waves can be a single crest or trough, or a series of alternating crests and troughs that radiate out from the center of the atom.

The hydrogen atom has just one electron, which occupies a spherical wave made up of a single crest or trough. The lithium atom has three electrons, one of which is outermost and corresponds to the peel of the atom-fruit. As in the hydrogen atom, this outermost electron occupies a spherical wave, but this time the wave has a crest and a trough. In the same way, the outermost of the eleven electrons of the sodium atom has a spherical wave with two crests and a trough.

Atom-fruits with similar peels tend to have similar chemical properties. This is particularly true for the number and direction of bonds they can form. Thanks to the spherical shape of their electron waves, atoms like hydrogen, lithium, and sodium can form bonds to other atoms in any direction. Since the crests of the waves are symmetrical, all directions are equally good.

(a)

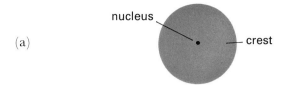

(b)

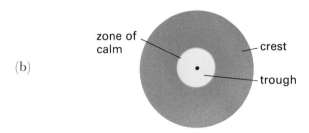

(c)

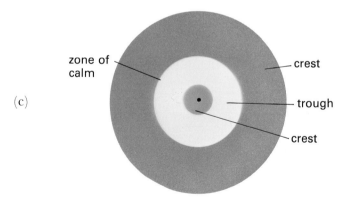

These cross sections of spherical stationary waves show different numbers of alternating crests and troughs.

a) For the unique electron of the hydrogen atom

b) for the third electron of the lithium atom

c) for the eleventh electron of the sodium atom

3 WAVES THAT LOOK LIKE FIGURE EIGHTS AND CLOVERLEAVES

As the number of electrons in an atom increases, they become more crowded. Fortunately, there are stationary waves that give the electrons more space. One has a cross section that looks like a figure eight. On one side the wave swells out into a crest, and on the other it sinks into a trough. The two lobes are separated by a zone of calm located precisely at the nucleus of the atom. Waves that look like figure eights are unique to the atomic world and cannot be found anywhere else. Electrons in an atom can form three of them, oriented in three perpendicular directions: east-west, north-south, and up-down. Each of these waves therefore has a specific orientation defined by the direction in which the crest and trough point. It is in these specific directions that bonds to other atoms can be formed. Important atoms like carbon, nitrogen, and oxygen have three figure eight waves in addition to the spherical waves discussed earlier.

An even more remarkable electron wave has a cross section that looks like a cloverleaf. The wave swells out into a crest in two of the four lobes and sinks into a trough in the other two. There are four waves of this type, identical except for their orientation in space. A fifth wave belongs to the same family but has a distinctly different shape that resembles two eggs sticking out of the hole in a doughnut. Members of the cloverleaf family of waves are commonly found in metal atoms like those of chromium, iron, and nickel. Because these waves have crests and troughs oriented in multiple directions, they allow metal atoms to form bonds to a large number of neighboring atoms. A single cloverleaf wave, for example, allows four bonds to be formed in a plane.

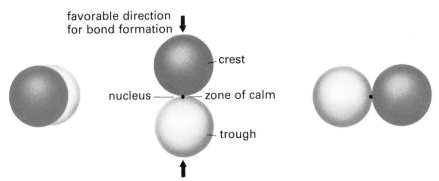

The crests and troughs of figure eight waves define the directions in which bonds can be formed. Atoms of carbon, nitrogen, and oxygen use three of these waves.

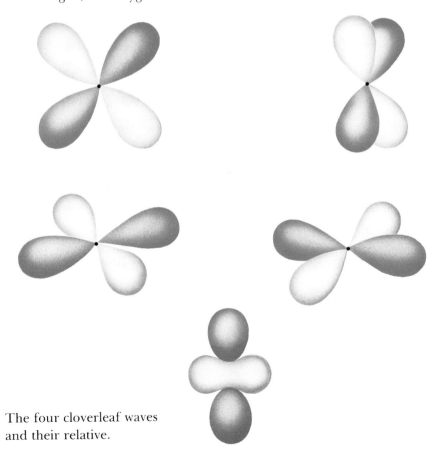

The four cloverleaf waves and their relative.

4

THE MATING GAME OF ELECTRONS

Each atom has a characteristic number of electrons. Hydrogen, the smallest atom, has only one, while the largest atoms have more than one hundred. The uranium atom, for example, has ninety-two electrons. This means that there are about one hundred different atoms. The nucleus of each is unique, and contains a positive electric charge that exactly balances the negative charge of the surrounding electrons.

Two electrons, but not more than two, can pair up to form and occupy a single wave. However, two electrons chosen at random do not necessarily make a happy couple. Each electron has a distinctive, intimate trait (called spin, not sex!) and must find a complementary mate. Paired electrons in a wave are represented by two arrows pointing in opposite directions. A solitary electron is represented by a single arrow.

The hydrogen atom has a single electron in a spherical wave, whereas the helium atom has a pair of electrons in a similar spherical wave. More complex atoms have electron pairs in other types of waves. The waves that are available are filled up with electrons in a very systematic way. Atoms are a little like onions, since the electron waves are organized in layers. Whether shaped like spheres, figure eights, or cloverleaves, certain groups of waves are equally accessible to electrons and can be considered to form a layer or shell. Within each shell, pairs of electrons occupy particular waves, represented by segments of the shell. Waves that are not occupied by electrons are considered empty. In a sense, these waves exist even if they contain no electrons, but they play no chemical role unless they are occupied by one or two electrons.

Atoms with shells that are completely filled with electrons are particularly stable. Two atoms whose outermost shells contain the same number of electrons have similar properties. This is why lithium and sodium atoms behave similarly. It is also the basis for the periodic table, an atomic classification scheme first proposed by the Russian chemist Mendeleev.

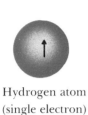

Hydrogen atom
(single electron)

Helium atom (pair of electrons)

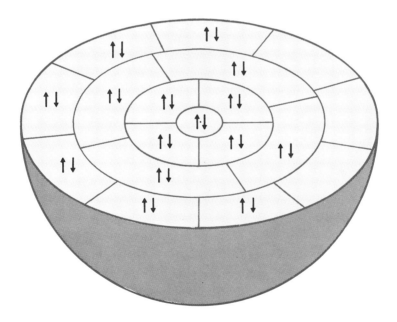

This drawing symbolizes how atoms are filled up with electrons shell-by-shell.

5

WHEN WAVE MEETS WAVE: FRIENDSHIP AND HOSTILITY

In order for two atoms to bond together and form a molecule or a fragment of a molecule, their waves must meet and interact. This encounter of waves belonging to two different atoms is at the heart of all chemistry.

Let's suppose that two waves come together and overlap. The result is two new waves that spread out over both atoms. In one, two crests meet, reinforce each other, and merge, producing a new wave with a single crest in place of two separate crests. This new molecular wave is very attractive to electrons. An electron in the wave is simultaneously attracted to the nuclei of both atoms, so it is doubly stabilized. This wave therefore helps hold the atoms together and is called *friendly*.

But woe betide electrons that must occupy the second molecular wave. This wave is formed when the crest of one atomic wave meets the trough of the other and cancellation occurs, producing a molecular wave in which a crest is separated from a trough by a zone of calm. Waves like this are nothing new, since we have already seen that figure eight and cloverleaf waves have crests and troughs separated by zones of calm. In the present case, however, the zone of calm lies right between the nuclei in the region where a bond should be forming. This makes the wave very unattractive to electrons, and they avoid it as much as possible. This wave does not help hold the atoms together and is called *hostile*.

The formation of friendly and hostile waves is similar to the effect of throwing two pebbles into a pond. The little concentric ripples around each impact point spread out and collide, leading to reinforcement (friendly wave) or cancellation (hostile wave).

The meeting of two waves . . .

hostile

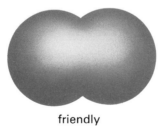

friendly

. . . produces one friendly wave and one hostile wave. The wave with a crest, trough, and zone of calm is hostile because the zone of calm lies between the two atoms.

6

THE TIES THAT BIND

Overlapping of the waves of two approaching atoms is especially favorable when the atoms contain solitary electrons with complementary spins. Interaction of the atoms allows the electrons to be paired up in the friendly molecular wave, and each electron is happier there than it was alone in an atomic wave. In this way, the hydrogen molecule serves as matchmaker for two electrons that were initially all alone in two separate hydrogen atoms. The stationary wave of this electron pair extends over both nuclei. This is the origin of the chemical bond between two atoms. Electrons serve as the glue that holds the atoms together. This idea, the fundamental concept of chemistry, is due to the American chemist Gilbert Lewis.

On the other hand, formation of a molecule from two atoms is forbidden when the electrons would have to occupy a hostile wave. Helium atoms have two electrons, so when two helium atoms interact, a total of four electrons are going to want to form molecular waves. One pair could occupy a friendly wave similar to that of the hydrogen molecule, but the other pair would be obliged to occupy a hostile wave. The second pair would be so unhappy that the molecule would not be able to survive. As a result, helium atoms take pains to avoid one another.

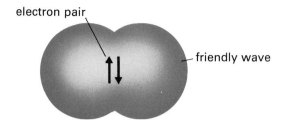

In the hydrogen molecule, an electron pair in the friendly molecular wave holds the atoms together.

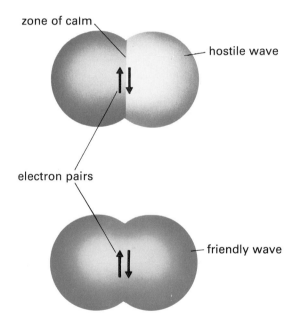

Formation of the helium molecule is forbidden because one of the two electron pairs would have to occupy a hostile wave with a zone of calm between the two atoms.

7

PETAL-SHAPED WAVES

Friendly waves occupied by pairs of electrons hold atoms together in molecules. The bonds are strongest when the atomic waves from which they are formed point toward each other and overlap well. To improve the overlap, an atom can blend its original waves and create new atomic waves specially directed toward neighboring atoms. These waves were discovered by the American scientist Linus Pauling.

For example, by mixing its three figure eight waves and one spherical wave, a carbon atom can produce four petal-shaped waves that point in four different directions in space. Each petal is made up of a large crest and a small trough. The four petal waves can interact with the spherical waves of four atoms of hydrogen to form four friendly molecular waves. Since the carbon atom contributes four electrons and each of the four hydrogen atoms contributes one, each of the friendly molecular waves will be occupied by an electron pair, creating four carbon-hydrogen bonds. As we have seen, the methane molecule that results has the shape of a tetrahedon. The ability of carbon atoms to form four bonds in a tetrahedral direction in space was discovered more than a hundred years ago by the Dutch scientist van't Hoff and the French scientist LeBel. It explains why left-handed and right-handed molecules exist, and it is the foundation of a field of chemistry called *stereochemistry*.

Some atoms, like those of nitrogen and oxygen, ordinarily use one or more of their petal waves to accommodate an electron pair called a "lone pair" because it is not necessarily used to form bonds. For these atoms, the power of attachment to other atoms, which is the famous *valence* of chemists, is therefore smaller than it is for carbon atoms. Unlike carbon, oxygen normally forms only two bonds and nitrogen only three.

The four petal-shaped waves of the carbon atom, with one shown in detail at the right.

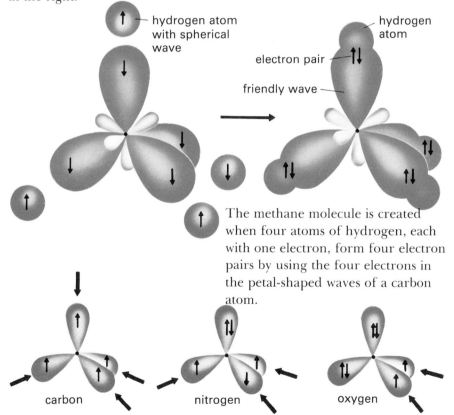

The methane molecule is created when four atoms of hydrogen, each with one electron, form four electron pairs by using the four electrons in the petal-shaped waves of a carbon atom.

The number of single electrons in petal-shaped waves (four for carbon, three for nitrogen, and two for oxygen) determines the power of attachment to other atoms (usually four bonds for carbon, three for nitrogen, and two for oxygen). The larger arrows show the directions in which bonds can be formed.

8

DOUBLE BONDS

Sometimes a carbon atom forms two or three bonds with one other atom. These double or triple bonds are represented by two or three lines drawn between the atoms involved. In the ethylene molecule, for example, two carbon atoms are doubly bonded to one another. Each carbon atom has three neighbors: two hydrogen atoms and one carbon atom. To form these three bonds, each carbon atom blends two of its three figure eight waves with its single spherical wave to produce three petal waves in the plane of the molecule. Each carbon atom therefore retains one unchanged figure eight wave that is oriented perpendicular to the molecular plane.

When the two carbon atoms interact, a petal wave contributed by one will overlap a petal wave of the other. This creates an inner friendly wave that includes both atoms and holds one electron pair. A second, outer friendly wave is formed at the same time by bringing together the two intact figure eight waves. On one side of the molecular plane, the two crests merge into a single crest, and on the other the two troughs merge into a single trough*. This creates an outer molecular wave that lies over and under the inner molecular wave like the bun covering a hot dog. The outer wave also holds a pair of electrons. Although the drawing shows both electrons in the crest of the wave, they can equally well both be in the trough, or one can be in the crest and one in the trough. The carbon atoms are therefore held together by a double bond, and they are closer together than carbon atoms joined only by a single bond.

The strength of a double bond gives a molecule special stability. Paradoxically, a double bond also makes the molecule more reactive, since the two carbon atoms can sacrifice one of the two bonds without having to separate completely.

---

*This outer wave therefore has a zone of calm in the plane of the molecule. However, this zone does not separate the two nuclei, unlike the zone of calm in hostile waves. Electrons in the outer wave are simultaneously attracted to both nuclei and are therefore perfectly happy.

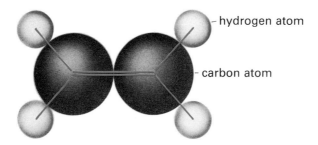

The ethylene molecule viewed from above the plane of the atoms.

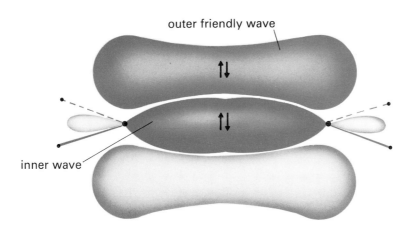

Side view of the two friendly waves of the double bond.

4
Over Hill, Over Dale

REACTING MOLECULES

1

THE DAILY LIFE OF MOLECULES

Whether in the test tubes of a chemist or in the air, water, or earth of our environment, molecules are always ready to react. Contact with a second molecule can lead to the formation of one or more new molecules by the creation of new bonds between the atoms of the molecules that collide. Sometimes these encounters do nothing but give one molecule enough energy to reorganize its bonds all by itself.

As in heating a house or starting a car, energy must ordinarily be provided to make molecules react. This energy allows the molecules to cross over the hardest part of the path that converts them into products. A molecule is like a bicycle racer who starts in a valley, climbs, crosses a pass, then descends into another valley.

To make the analogy perfect, though, we would have to imagine that the cyclist somehow arrives at the destination completely transformed. This is because in every reaction, radical changes occur in the positions of the atoms, and the molecular skeleton is fundamentally altered.

Reactions have been classified by chemists according to the kind of structural change that takes place. The type and ease of reaction depend critically on the fate of the electron pairs in the reacting molecules. Sometimes an electron pair stays together, and the wave that contains them simply changes its shape. Sometimes the electrons in a pair split up and go their separate ways, each in its own wave. Finally, a new pair is sometimes formed from two single electrons.

Like a cyclist who must cross a pass, a molecule needs energy in order to react. This is also true when several molecules react together.

2

UMBRELLAS, CHAIRS, AND BOATS

The simplest reactions are those in which a molecule singlehandedly turns into something else. Sometimes these changes are only temporary, and the molecule returns to its original state.

In the ammonia molecule, the nitrogen atom uses three petal waves to form bonds to three hydrogen atoms. The hydrogen atoms define three of the four corners of a tetrahedron surrounding the central nitrogen atom. The fourth petal wave of the nitrogen atom points toward the fourth corner and, as we have seen, holds a lone pair of electrons not involved in bonding. The ammonia molecule therefore resembles an umbrella. Like an umbrella, it can open up and turn itself inside out. This reaction is very fast, and the molecule inverts more than ten billion times per second. The inversion is accompanied by a change in direction of the wave carrying the lone pair of the nitrogen atom, but the paired electrons are not split up.

The cyclohexane molecule has a framework of six carbon atoms in a ring. The carbon atoms are linked by single bonds, with one electron pair per bond. At rest, the cyclohexane molecule looks like a chair, but it is very flexible. If we add a little energy, the chair distorts and assumes various shapes. First it becomes a sofa, then a boat. In the course of these distortions, the electron pairs that hold the carbon atoms together wander a bit, but they stay together. Like pearls in necklace, the carbon atoms are free to move, but they remain held together by a string of electron pairs.

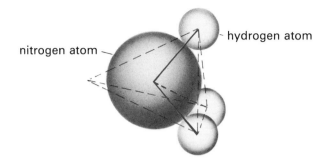

In the ammonia molecule, three hydrogen atoms and a lone pair of electrons define the four corners of a tetrahedron surrounding the central nitrogen atom.

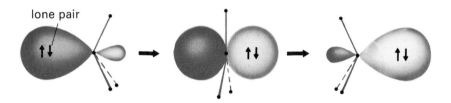

When the ammonia umbrella-molecule turns inside out, the lone pair stays intact.

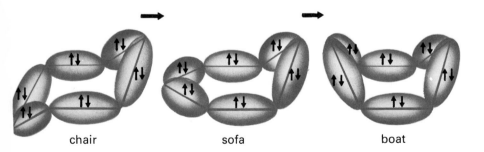

chair   sofa   boat

When the skeleton of six carbon atoms in the cyclohexane molecule distorts, the electron pairs stay together.

3

ONE PAIR KICKS OUT ANOTHER

Now let's consider reactions in which two molecules interact and turn into something new. In most of these reactions, electron pairs stay together. Let's take a molecule of hydrochloric acid, in which a hydrogen atom and a chlorine atom share an electron pair in a friendly wave, and let's bring it toward an ammonia molecule, which we already know has a lone pair in a petal wave. When these two molecules come together, they dock like a spacecraft and a lunar landing module. During their brief contact, the hydrogen atom of the hydrochloric acid molecule moves to the ammonia molecule, but abandons its electron on the chlorine atom that has been left behind. In this way, the electron pair that held the hydrogen and chlorine atoms together and the free electron pair of the ammonia molecule both remain intact. A negatively charged chlorine ion and a positively charged ammonium molecule are formed. In this reaction, a hydrogen atom is literally torn away from one electron pair by another.

Now let's take a molecule related to methane, but with a fluorine atom at one vertex and three different unspecified atoms at the other three vertices. When a negatively charged fluorine ion approaches this molecule, it can try to form a bond to the central carbon atom. Since this atom already has the maximum number of four bonds, one of the atoms to which it is attached must leave to allow a new one to take its place. It is the fluorine atom originally attached to carbon that leaves.

A special feature of this reaction is that as the invading fluorine atom approaches one side of the molecule, the fluorine atom that is being kicked out leaves from the other side. As a result, the umbrella formed by the other three unspecified atoms is turned inside out. Like the lamps with three different feet, the starting molecule and the product are not identical, but form a left- and right-handed pair.

In this substitution reaction, one electron pair kicks out another. The attacking fluorine ion carries an electron pair in a figure eight wave that is used to form the new carbon-fluorine bond, and the departing fluorine atom takes with it the pair that served to form the original carbon-fluorine bond.

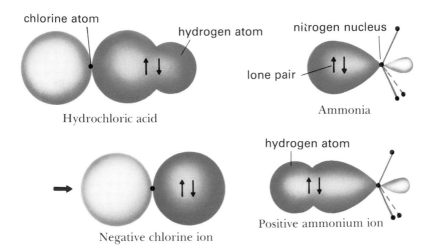

In most reactions, electron pairs stay together. The only waves of the chlorine atom and the ammonia molecule that are shown in detail are those involved in the reaction.

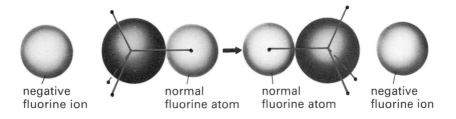

In a substitution reaction, one electron pair kicks out another.

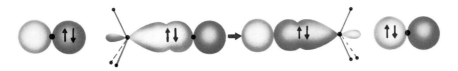

4

BREAKING UP IS HARD TO DO

Let's try to break up a pair of electrons. One of the easiest to split up is the one in the outer friendly wave of the carbon-carbon double bond in the ethylene molecule. Simply pulling the carbon atoms apart is not the easiest way to split up this pair, since we would be fighting against two pairs at once, one in the inner wave and one in the outer. Instead, an easier way is to twist the molecule by rotating the plane of the three atoms on the right relative to the plane of the three atoms on the left. By twisting the molecule in this way, we reorient the figure eight waves of the two carbon atoms, which were initially parallel in the ethylene molecule. This destroys the outer friendly wave, since the crests and troughs of the figure eight waves no longer overlap, but it does not harm the inner friendly wave.

After a rotation of 90 degrees, the rupture of the outer electron pair is complete, and the two ends of the molecule are held together by only one pair of electrons. The twisted molecule is very unstable. The rotation tends to continue until, at 180 degrees, the two hydrogen atoms at one end of the molecule have exchanged their original positions. At this point, the two separated electrons find each other again and are reunited as a couple. The ethylene molecule is then restored.

Breaking up even this outer electron pair requires an impressive amount of energy, and the molecule must be heated to over 500°C. To confirm that the twisting has actually taken place, a chemist can label the ethylene molecule in advance with two atoms of deuterium. Deuterium atoms have one electron and waves similar in every respect to those of hydrogen atoms, but deuterium atoms have heavier nuclei. By labeling the ethylene molecule with these *isotopes* of hydrogen, we can distinguish the starting material from the product.

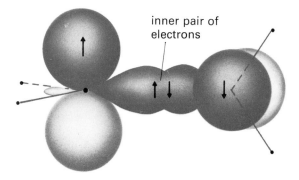

inner pair of
electrons

Separation of the outer pair of electrons in a twisted
ethylene molecule.

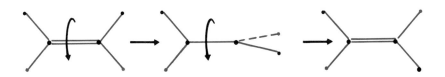

Deuterium isotopes, represented by blue nuclei, make it
possible to show that the molecule has been twisted.

5

FIDELITY AND THE ETERNAL TRIANGLE

Breaking up an electron pair, whether temporarily or permanently, is usually even harder than it is for the outer pair of the ethylene molecule. Molecular divorce is uncommon. We can even say that molecular reactions are governed by an eleventh commandment:

Thou shalt not split up a pair of electrons.

The pairs that make up single carbon-hydrogen and carbon-carbon bonds are among the most faithful and difficult to break up, which accounts for the stability of molecules like methane. The electron pair that makes up each bond is solidly planted in a very friendly wave and blissfully happy. Drastic heating may break up the pair, but the whole molecule will probably be torn apart, too. Breaking up the happy couple without destroying the household requires more subtle means.

A seemingly radical way to split up the pair uses a third electron, a bachelor in a second molecule. When such an electron meets the pair, things happen fast. As sometimes happens in daily life, the ardent bachelor tries to take what interests him, the electron with the complementary spin. If the bachelor succeeds and replaces the lawful mate, which is often the case, a bond breaks in the molecule under attack and a new bond forms in the intruder. In this way, a chlorine atom can take a hydrogen atom away from a methane molecule to form a molecule of hydrochloric acid.

However, it is important to note that a new pair of electrons in the intruder appears in place of the original pair. Since electrons are all the same and cannot be distinguished from one another, it is possible to say that the original pair of electrons has merely traded places with the single electron. In this sense, the commandment has not really been violated.

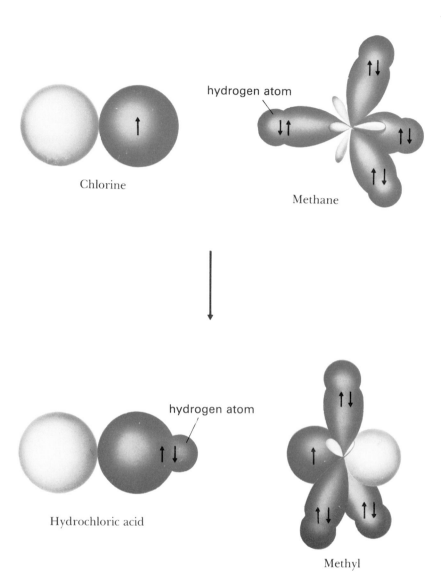

Chlorine

hydrogen atom

Methane

Hydrochloric acid

hydrogen atom

Methyl

The bachelor electron of a chlorine atom steals a mate from the pair that makes up a carbon-hydrogen bond, and in this way the chlorine atom pulls off a hydrogen atom.

6

ELECTROCHEMISTRY

In electrochemistry, a technique discovered by the English scientist Faraday, electrons are removed from or added to a molecule. The molecule is brought up to an electrode, a piece of metal that carries an electric charge. The electrode can be short of electrons, positively charged, and eager to take them from the molecule; or it can be overloaded with electrons, negatively charged, and eager to transfer them to the molecule.

Like the methane molecule, the acetone molecule contains bonds in which the electron pairs are well entrenched. This is particularly true for the two carbon-carbon bonds. However, if an acetone molecule is brought up to an electrode that wants electrons badly, an electron will leave the molecule. If enough energy is provided, the electron may even come from one of the pairs that form the carbon-carbon bonds, leaving a single electron in the friendly wave of one of the carbon-carbon bonds. This bond will then be rather easy to break, producing an acyl molecule, which has a single electron, and a positive methyl ion*.

In other molecules, it is possible to split up a pair that forms a bond by adding a third electron from an electrode with electrons to spare. This electron enters a hostile wave of the bond under attack. As we have already seen in the case of two helium atoms, the presence of an electron in the hostile wave destabilizes the bond. The bond breaks, and the electron pair splits up.

Electrochemistry is therefore a powerful industrial tool for creating new molecules. Molecules can be broken apart and then reassembled in more useful forms. For example, electrochemical synthesis can be used to make the nylon molecule discussed earlier.

---

*Acetone molecules do not actually undergo this plausible reaction, but closely related molecules do.

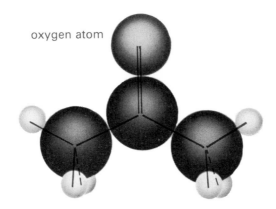

The acetone molecule

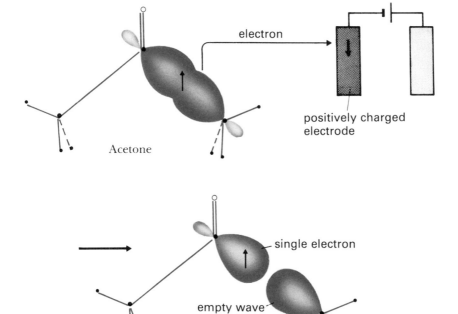

Having lost one of its electrons to an electrode, the carbon-carbon bond breaks and an acyl molecule and a positive methyl ion are formed.

PHOTOCHEMISTRY

7

Everyone has seen inks fade in the sun and newspaper turn yellow when it is exposed to light for a long time. These changes are the result of reactions caused by light, called photochemical reactions. Light lets electrons behave in an uncharacteristic way. It provides the energy needed to break up a pair of electrons by giving one of them a kick. This electron is suddenly knocked into a hostile wave located in a different part of the molecule.

Let's take an acetone molecule and shine light on it for a moment. One of the electrons in the carbon-carbon bond can be kicked into a hostile wave of the carbon-oxygen bond. The carbon-carbon bond, now held together by only one electron, is likely to break.

From the energetic point of view, photochemical reactions are remarkable. Light miraculously changes the path that a reacting molecule must follow. Like a cyclist lifted up to a peak somewhere high above the pass, the molecule no longer faces an arduous climb. Instead, it swoops straight down the slope and is easily converted into products. In this way, light acts like a cable car that lifts molecules from low-energy starting points to high-energy summits. On the descent, they undergo transformations easily.

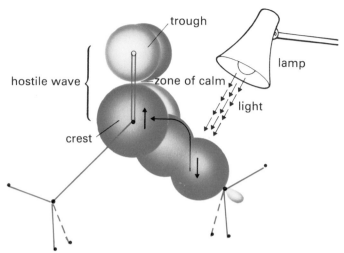

Light kicks an electron from a friendly wave of the carbon-carbon bond into a hostile wave of the carbon-oxygen bond. The carbon-carbon bond, now held together by only one electron, is likely to break.

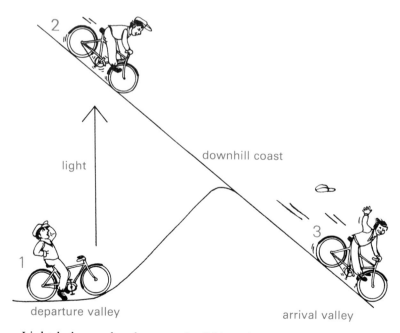

Light helps molecules react by lifting them up to high-energy summits.

8 CATALYSIS AND THE MAGIC OF SURFACES

It would be nice to be able to break the bond between two atoms without having to split up the electron pair that forms the bond. Something like this can be done by certain molecules that have a large metal atom at their center. The metal atom, like the rhodium atom with four bonds in chlorotris(triphenylphosphine)-rhodium, cleverly uses its many cloverleaf waves to break up approaching molecules. For example, if a hydrogen molecule comes close, the two hydrogen atoms will separate and each will form a new bond to the rhodium atom, which will then have a total of six neighbors. The new molecule is chemically very reactive since rhodium atoms normally prefer to form only four bonds. As a result, the hydrogen atoms are easily transferred to other molecules. This process, called hydrogenation, is widely used in industry to improve the quality of edible oils.

Let's stack up iron atoms until we get a speck of iron that is not big enough to be seen by the naked eye, like a piece of copper wire, but still big enough to contain thousands and thousands of atoms. On the surface of this microscopic metallic particle, the iron atoms are neatly lined up. Now let's expose the particle to nitrogen gas, which is made up of molecules of nitrogen. These molecules zip around and occasionally approach the particle of iron. When the nitrogen molecules hit the surface, they stick and split up into atoms. The three bonds that joined the two atoms of nitrogen are broken, and each atom becomes attached to an atom of iron on the surface of the particle.

If hydrogen molecules are mixed with the nitrogen molecules, both will stick to the surface and split up into atoms. The hydrogen and nitrogen atoms can then combine to produce molecules of ammonia. This enormously important reaction is used for the industrial production of ammonia molecules, from which many cleaners and fertilizers are made. The iron acts as a *catalyst* by cleverly facilitating a reaction of hydrogen and nitrogen molecules that would ordinarily be extremely slow. The detailed mechanism of this catalysis is still a mystery.

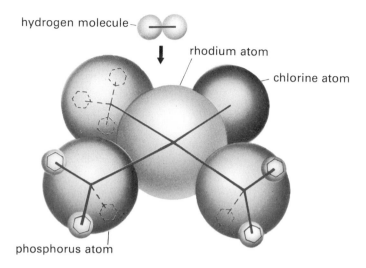

Hydrogen molecules can be split up by catalysts like the chlorotris(triphenylphosphine)rhodium molecule. Each little hexagon represents a phenyl group, which is similar to a benzene molecule. One of the six carbon atoms of the phenyl group is bonded to a phosphorus atom.

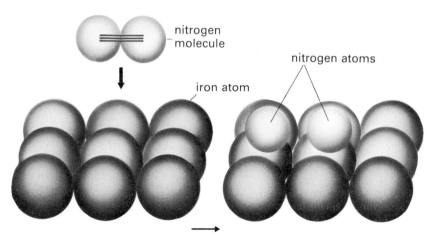

An iron surface can easily break all three bonds of a nitrogen molecule.

5

To React or
Not to React

THE AMAZING POWER OF ELECTRONS
TO SAY YES OR NO

1

A YES OR A NO TO A CREST OR A TROUGH

Stationary electron waves do not just make pretty pictures. The patterns of crests and troughs in these waves actually help control the very shape of molecules, their reactions, and therefore all the world around us. The wide variety of chemical reactions reflects the complex nature of electron waves.

The three figure eight waves of the carbon atom, the predominant atom in organic molecules, nicely illustrate the complex ways in which waves interact. For example, let's consider the following question. Can two figure eight waves, joined directly together in a friendly wave like the outer wave of a double bond, have a friendly interaction with a figure eight wave provided by a third atom? The answer is *yes* if the third wave orients its crest toward the combined crests of the other waves; but *no* if it directs its crest toward one figure eight wave and its trough toward the other. In the second case, cancellation would occur on one side, and a zone of calm would be created between two of the atoms. This would lead to the formation of a wave with hostile character on one side, which would counterbalance the friendly character on the other. This example shows how important the exact orientation of reactants is in determining whether or not a reaction will take place.

It is also possible to ask if a chain of figure eight waves can be closed into a ring by direct friendly overlap of the waves at the ends of the chain*. Again, the answer is yes or no. In this case, as the figure shows, the answer depends on the number of waves in the chain. This suggests that even within a very closely related series of molecules, extreme differences in behavior may be observed.

*The need to close the chain of waves by friendly overlap will be discussed later.

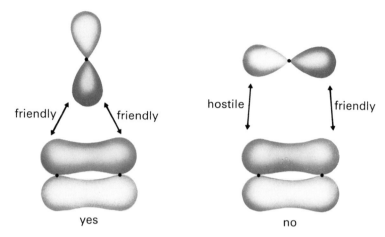

friendly friendly hostile friendly

yes no

Can two figure eight waves that are joined directly in a
friendly wave have a friendly interaction with a third wave?
The answer is yes if reinforcement occurs on both sides, but
no if cancellation occurs even on just one side.

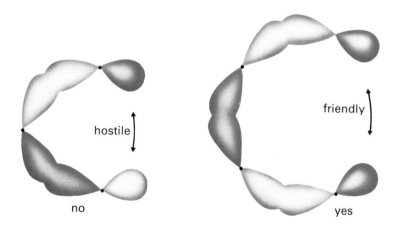

hostile friendly

no yes

Can a chain of figure eight waves be closed into a ring by
friendly overlap of the waves at the ends of the chain? The
answer depends on the number of waves in the chain.

EXTENDED WAVES

In the benzene molecule, each carbon atom has three neighbors, including one hydrogen atom and two carbon atoms. Since hydrogen atoms can form one bond and carbon atoms can form four, a double bond must link each carbon atom to one of its two neighboring carbon atoms. The molecule would appear to have to choose between the two structures shown in the diagram, as the German chemist Kékulé first suggested in 1867. In fact, the molecule cannot come to a decision. This hesitation between two possible structures, called *resonance* by the American chemist Pauling, is a stabilizing factor. The molecule actually adopts an intermediate structure in which the carbon-carbon bonds are all equal in length, neither fully double nor fully single.

This special geometry results directly from the shape of the friendly waves of the benzene molecule, which are put together like those of the ethylene molecule. First, the six carbon-carbon single bonds on the periphery are formed from petal waves. Six figure eight waves remain, one on each carbon atom. Their side-by-side interaction creates six waves that extend over the entire molecule. As in the simple interaction of two atomic waves, equal numbers of friendly waves (three) and hostile waves (three) are formed. Trace the crests and troughs of one of the friendly waves and notice how they extend over several atoms. In one of the friendly waves, the crest and trough actually make a full turn around the molecule. In the two others, the crests and troughs are divided into two regions by a zone of calm, and are therefore less extensive.

The wave with the complete loop is the most friendly. In general, the most friendly waves have the most extensive crests and troughs, and the fewest interruptions created by zones of calm between two atoms. In the benzene molecule, the other two friendly waves have more zones of calm and are therefore less friendly than the loop. Each of the three friendly waves will want to accept an electron pair. In principle, the three pairs could have been used to form three distinct double bonds in the molecule. Instead, they enter the three extended friendly waves and become evenly distributed around the periphery of the molecule. As a result, all six carbon-carbon bonds are equivalent.

The benzene molecule cannot choose between these two structures.

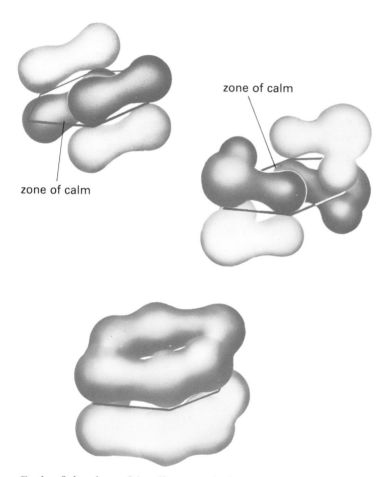

zone of calm

zone of calm

Each of the three friendly extended waves of the benzene molecule contains an electron pair.

3

THE 2,6,10 RULE

A cyclobutadiene molecule is similar to a benzene molecule, except that it has a periphery of four carbon atoms instead of six. Sought by chemists for decades, the cyclobutadiene molecule is a fugitive that can only be isolated briefly at very low temperatures. It can be stabilized by metal atoms in the form of supermolecules like cyclobutadieneiron tricarbonyl, which resembles the sandwich molecules seen earlier. How do we explain why the square cyclobutadiene molecule is so much less stable than the hexagonal benzene molecule?

This difference between two similar structures shows that very exacting laws govern the existence of molecules. These laws are enforced by electrons and their waves. One basic law governs the stability of molecules and determines whether or not a particular arrangement of atoms forms a stable structure. This law states that electron pairs should occupy friendly waves only. As we have seen, molecules of benzene, hydrogen, methane, and ethylene obey this law. But in the cyclobutadiene molecule, only one of the two electron pairs in extended waves is suitably located in a friendly wave. The other is forced to accept a new kind of wave that is neither fully friendly nor fully hostile. The cyclobutadiene molecule is unstable because this "indifferent" wave is occupied.

The striking difference between molecules of benzene and cyclobutadiene illustrates the 2,6,10 rule in chemistry. Molecules with extended waves formed by the side-by-side overlap of a continuous ring of figure eight atomic waves are stable only when they have 2,6,10, . . . electrons, which is equivalent to an odd number of electron pairs.

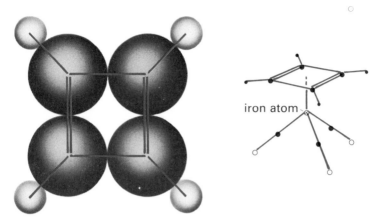

iron atom

The cyclobutadiene molecule is very unstable except when it is part of supermolecules like cyclobutadieneiron tricarbonyl, which is sketched on the right.

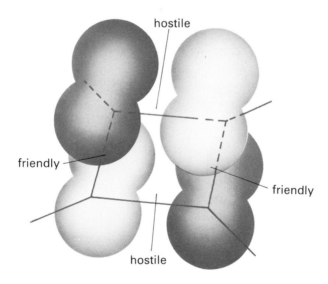

hostile

friendly

friendly

hostile

The instability of the cyclobutadiene molecule is due to the presence of an electron pair in an "indifferent" extended wave, which is neither fully friendly nor fully hostile.

4

FORBIDDEN AND ALLOWED REACTIONS

Let's bring a pair of ethylene molecules toward one another. Since the carbon atoms share two bonds and two electron pairs, each molecule could afford to give up one bond and one electron pair without falling apart completely. The two liberated pairs of electrons could then be used to form two new bonds *between* the molecules of ethylene. This would produce the well-known cyclobutane molecule. However, this reaction does not take place, even when ethylene molecules are heated strongly.

But now let's bring toward one of the ethylene molecules a molecule of butadiene, which has a framework of four carbon atoms with two double bonds and two electron pairs in extended waves. This time, with hardly any heating, the two molecules combine readily. The molecule formed, cyclohexene, has six carbon atoms and one double bond. This is an example of the Diels-Alder reaction, named for its German discoverers. Because this reaction creates rings of atoms easily, it is extremely valuable for the chemical synthesis of molecules.

The success of the Diels-Alder reaction, in which three electron pairs are used to form a ring of six atoms, is closely related to the stability of the benzene molecule, which is due to the presence of three electron pairs in a ring. In striking contrast, the analogous formation of a ring of four atoms is unsuccessful, and the cyclobutadiene molecule is unstable.

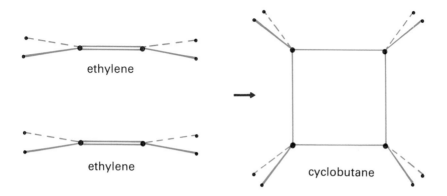

This reaction does not take place. When two ethylene molecules meet, they separate unchanged.

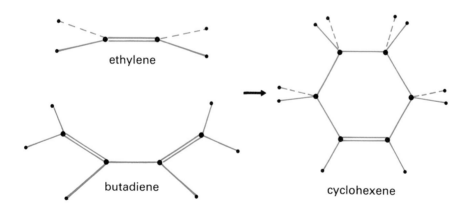

This reaction takes place readily. When an ethylene molecule meets a butadiene molecule, a new molecule of cyclohexene is formed.

5 | A LAW GOVERNING THE EVOLUTION OF WAVES

The American chemists Woodward and Hoffmann noticed the contrast between these forbidden and allowed reactions and were able to explain it by comparing the waves of the reactants and products. For the two preceding reactions, the comparisons are rather complicated. A simpler case is therefore illustrated here, a forbidden reaction in which two hydrogen molecules interact and try to form a single molecule with four hydrogen atoms. For each reactant, the right side of the wave occupied by an electron pair is identical to the left side. Such waves are called *symmetric*. But in one of the waves of the product, a crest on one side would face a trough on the other, with a zone of calm in between. Such waves are called *antisymmetric*.

The illustrated reaction of two hydrogen molecules would therefore require a profound change in the symmetry of one of the friendly waves. Such a change, accompanied by the sudden appearance of a zone of calm that was not initially present, is equivalent to splitting up an electron pair in the wave. Since our eleventh commandment prohibits these breakups, the reaction is forbidden.

Woodward and Hoffmann have shown that the symmetry of friendly waves must always be conserved in chemical reactions in order to avoid splitting up electron pairs.

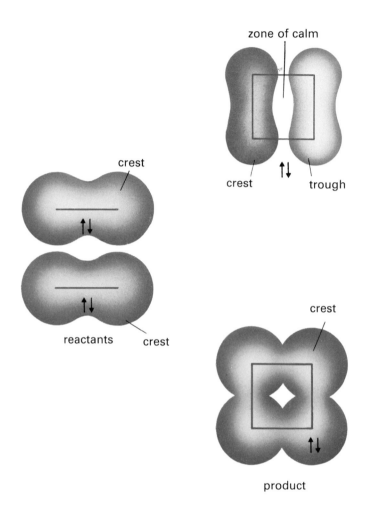

Comparison of the waves of the reactants and the hypothetical
product shows that the combination of two molecules of hydrogen
is forbidden. The waves of the reactants are symmetric, but one
of the waves of the hypothetical product is antisymmetric and
split by a zone of calm.

6

A SNAKE THAT BITES ITS OWN TAIL

Given enough energy, the butadiene molecule can react by itself and turn its chain of four carbon atoms into a ring. This produces a new molecule called cyclobutene. For this to occur, the pairs of hydrogen atoms at both ends of the chain must rotate so that the figure eight waves of the terminal carbon atoms are brought face to face. Then they can overlap and form a new friendly wave. Woodward and Hoffmann realized that this reaction can occur in two ways depending on whether the two ends rotate in the same direction or in different directions, and they showed that only one pathway is allowed.

According to the Japanese chemist Fukui, the course of a reaction is determined by the shape of the least friendly wave that contains an electron pair. In butadiene, this particular wave spreads out over the entire molecule like the extended waves of the benzene molecule. On either side of the molecular plane, it has a crest at one end of the chain facing a trough at the other. To have a friendly interaction between the ends of the chain, two crests obviously have to meet. For the molecular snake to bite its own tail, a crest at one end of the chain must meet a crest at the other, which requires that the two ends rotate in the same direction.

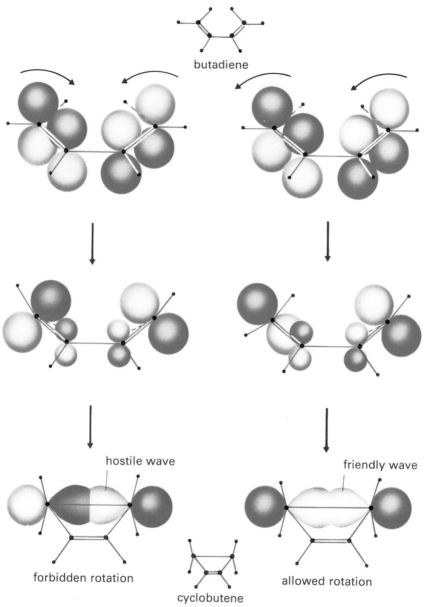

butadiene

hostile wave

friendly wave

forbidden rotation

allowed rotation

cyclobutene

The butadiene molecule can close to form the cyclobutene molecule in two conceivable ways. The actual outcome is determined by the evolution of the least friendly extended wave. (During the closure, the amplitude of this wave on the two central carbon atoms disappears).

7

BOXING AND MOLECULAR COLLISIONS

We have seen that a reacting molecule is like a cyclist who faces a road full of twists and turns. Just as cyclists tend to take a particular curve differently depending on their speed and skill, molecules follow different reaction pathways depending on how they are spinning and dancing around. These movements affect their response to the simple layout of the road.

An example is the encounter of a deuterium atom and a hydrogen molecule. The single electron of the atom will try to split up the electron pair of the molecule. As the atom approaches, the molecule continues its characteristic dance, and the bond between the hydrogen atoms lengthens and shortens rhythmically. If the atom surprises the molecule at full extension, a reaction occurs, and the atom bonds to one of the two atoms of the molecule. If the atom hits the molecule at full contraction, however, no reaction takes place.

This is like a boxer trying to hit his opponent. If he lands a blow when the adversary is backing up or dodging, the result is less telling than when he strikes while the opponent is advancing. In the same way, an efficient reaction requires more than just a suitable arrangement of electron waves. The instantaneous arrangement of atoms in the system must also be correct.

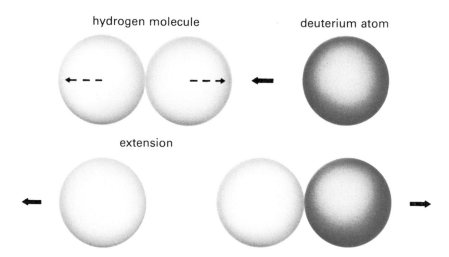

hydrogen molecule   deuterium atom

extension

A reaction takes place when the atom surprises the molecule at full extension.

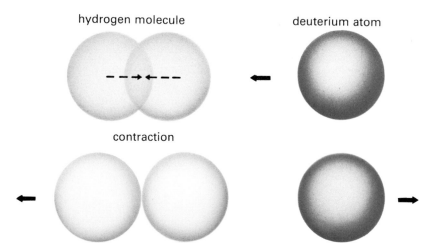

hydrogen molecule   deuterium atom

contraction

No reaction takes place when the atom hits the molecule at full contraction.

(Time advances from top to bottom).

8

THE MESSAGE OF MOLECULES

Molecules tell us about themselves in many ways. For example, atoms in isolated molecules spin and dance according to strict laws. At any moment, a molecule can change its motion by exchanging a tiny amount of energy with the outside environment. This exchange can be made to produce a recordable signal that provides important information about the molecule.

Such a signal is given when light shines on a molecule, breaks up an electron pair, and kicks one into a hostile wave. During this process, the molecule and other molecules like it absorb a tiny bit of light energy that corresponds to a characteristic color of light. The material formed from these molecules therefore appears colored to the eye. The material is white if the wave that the electron is kicked into is very hostile; yellow or orange if the wave is less hostile; and red, green, or blue if the wave is not very hostile. A tulip will be yellow if its pigment molecules have very hostile waves, but purple if the waves are only a little hostile.

Molecules therefore reveal themselves to us in countless ways, even if we cannot see them directly. Of these revelations, life is the masterwork. Some acts, like the pass of a football player or a mother's kiss, will always touch us in a special way. From now on, though, we may look at these acts with an inner smile of new appreciation for all the marvelous molecules that make them possible.

The petals of yellow tulips contain molecules with waves that are very hostile.

The petals of purple tulips contain molecules with waves that are not very hostile.

A SHORT GLOSSARY FOR CHEMISTS

Word Used in This Book	*Scientific Translation*
wave	orbital
friendly	bonding
hostile	antibonding
indifferent	nonbonding
spherical wave	s orbital
figure eight wave	p orbital
cloverleaf wave	d orbital
petal wave	sp^n hybrid
hydrogen bridge	hydrogen bond
zone of calm	node
crest	positive amplitude
trough	negative amplitude
2,6,10 rule	$4n + 2$ rule
suitcase	crystal